THE EDGE OF COHERENCE

A Strategic Treatise on Distributed Energy Sovereignty for the Twin Cities and Its Emerging Fringe

ISBN: 978-1-970547-12-2

A narrative corridor, a ciphered essay, and a technical white paper, assembled as a single document of intention and clarity.

Author's Note on Scope and Use

This work presents an analytical and conceptual exploration of system design, energy conversion, and applied infrastructure strategy. It is intended to provide clarity, structure, and perspective on emerging approaches, including SET smart gasification.

It is not a technical manual, engineering specification, or site-specific recommendation.

Any systems, technologies, or frameworks discussed are presented for informational purposes only. No representations or guarantees are made regarding performance, outcomes, or suitability for any particular application.

Nothing in this work should be interpreted as professional, engineering, environmental, or financial advice. Implementation of any system described requires independent evaluation, site-specific analysis, and appropriate technical and regulatory consultation.

References to technologies or systems developed by others do not imply authorship, endorsement, or validation by those parties unless explicitly stated.

FRONTISPIECE

There are moments when a region stops negotiating with its past and begins speaking directly to its future. Those moments rarely arrive with ceremony. They arrive as quiet recognitions — that the center has grown heavy, that the edge has grown capable, and that the world no longer bends to the architecture that once defined its sense of order.

This document was written for that moment.

AUTHOR'S CLARIFICATION

As you read this document, I want to offer a brief reflection that comes directly from my intent in writing it. Although this work references the Twin Cities—Minneapolis and St. Paul—it is not written for the core metro. It is written for the cities and counties on the fringe, the places that have carried the burdens of the region's waste and energy decisions without receiving commensurate infrastructure in return.

When I speak of the Twin Cities here, I am acknowledging their existing legacy systems: the incinerators, the district energy loops, the centralized thermal assets that have defined urban utility for several decades. These facilities will not vanish, nor am I suggesting that they should. They will continue to operate as the backbone of the metropolitan core.

But the real opportunity—the market I care about, the communities I am addressing—is the belt of cities, counties, tribal nations, and industrial corridors that surround the metro. The first ring, the second ring, the outlying towns, the rural-industrial margins, the airport-adjacent land, the tribal sovereignties at the edge. These are the places where need outpaces legacy infrastructure, where waste accumulates without technical relief, where PFAS, plastics, agricultural residues, and stranded streams pose challenges the metropolis is not structured to solve.

This document was built for them.

The Twin Cities are my reference point, not my target. They provide the contrast. The outlying regions provide the opportunity.

I am not asking the metro to dismantle its incinerators. I am not replacing HERC. I am not critiquing District Energy St. Paul. I am simply acknowledging that their geography has limits—and that the future of resilience, sovereignty, and distributed carbon precision does not lie in the center, but in the periphery.

The SET Smart Gasification platform is the tool built for that periphery: modular, flexible, unconcerned with density, capable of serving places that legacy systems cannot reach and were never meant to reach. My goal is not to confront the metro's infrastructure, but to complete it by filling the gaps the core cannot reasonably address.

The counties and cities beyond the beltway deserve the same energy sophistication, the same carbon intelligence, and the same agency that the core inherited decades ago. With SET, they no longer need to wait for permission from the center to solve problems created outside their control.

So although this document speaks of the Twin Cities, it is the outlying areas—the fringe, the forgotten corridors, the sovereign lands, the rural-industrial seams—that move me to write. This work belongs to them.

FRINGE-CITY ABSTRACT

The region's future will not be shaped in the towers of the Twin Cities nor in the legacy assets that heat their streets. It will be shaped in the ring of municipalities, counties, tribal lands, industrial corridors, and rural-urban seams that surround the metro—places that have long carried the burdens of the center without receiving the infrastructure that matches their responsibility. This document was written for those communities. The centralized thermal systems of Minneapolis and St. Paul will continue to serve the urban core with honor and longevity, but they were never designed for the geography beyond their reach. The outlying cities face a different landscape entirely: rising hauling costs, PFAS liabilities, stranded waste streams, agricultural residues, brownfield constraints, local political pressure, and the expectation to do more with less every year. SET Smart Gasification enters this environment not as a replacement for the metro's incinerators, but as the first technology built specifically for the edge—modular, precise, indifferent to density, and capable of operating where centralized infrastructure no longer applies. For fringe communities, this means the possibility of local control over waste, local control over carbon, and local control over the economic and environmental consequences of decisions made far from their borders. It means eliminating dependence without rejecting the metro. It means complementing the center while finally claiming the sovereignty the edge has earned. This abstract exists to state one truth plainly: the regions beyond the beltway are no longer peripheral. They are the frontier of a new energy order. Everything that follows—Prelude, Essay, White Paper—is written to confirm that future.

EXECUTIVE INTRODUCTION FOR SUBURBAN/EXURBAN LEADERS

The cities beyond the first ring have always lived in a quiet paradox: close enough to feel the gravitational pull of the metropolitan core, far enough to bear the weight of its unmet obligations. These are the communities that serve as buffers, corridors, industrial engines, and residential havens, yet they are rarely acknowledged as the places where the future's first pressure points appear. This introduction exists to correct that oversight.

Suburban and exurban leaders stand at the threshold of a transformation that no longer depends on downtown infrastructure to validate itself. The geometry of energy and waste is shifting outward, following the logic of land availability, industrial distribution, rising population, and the chemistry of modern materials. Where the center once commanded, the perimeter now decides. And the decisions made in these municipalities — in transfer stations, industrial parks, growing districts, and sovereign plots of tribal land — will shape the regional order far more than any legacy facility in Minneapolis or St. Paul. SET Smart Gasification is not an urban technology scaled down. It is a perimeter technology scaled precisely. It honors the agency of the suburban and exurban corridor by offering a future in which these communities are not the recipients of metropolitan overflow, but the stewards of their own resource intelligence. This document was shaped with that reality in mind: the next chapter of the region will be written where the map thins and the responsibilities thicken.

To those leaders, this introduction stands as recognition — and invitation.

PRELUDE

There are documents that arrive as summaries, and there are documents that arrive as rehearsals. Then there are those rare constructions that refuse the conventions entirely—documents that appear on the desk like quiet emissaries of a shift already underway. They do not request attention. They assume it. They carry the unmistakable posture of something that has outgrown the limitations of a single form.

If you are reading these pages, you have entered such a document.

It will not tell you where it begins, because the beginning is no longer the most honest place to start. The heart of the matter lies not in chronology but in recognition—an understanding that the world has rearranged itself quietly, while our older infrastructures continued to operate as if the century were still taking its instructions from them.

But a region cannot negotiate with time by insisting it remain obedient. It must acknowledge when the future has slipped past the perimeter and is waiting, with almost indecent patience, for the present to notice.

This Prelude is the noticing.

It exists to prepare you—not with data or arguments, but with atmosphere. You are not about to read a white paper. Nor are you merely reading an essay. What follows is a corridor connecting two ways of knowing: the narrative intelligence that reveals why a system must change, and the technical intelligence that shows how it does.

The Twin Cities, in their own way, have always understood these dualities. The city that builds a steam loop beneath its streets and an incinerator on its riverfront is a city fluent in both poetry and pragmatism. It knows that winter demands structure and that progress demands narrative. What it has not admitted—at least not yet—is that the structures which once defined its competence are beginning to hum out of tune with the century that surrounds them.

Cities age not by cracking but by continuing too long in the belief that their purpose does not need revision. Waste arrives differently now. Carbon behaves differently now. Public patience has reconfigured itself into an instrument far more precise than the politics that claim to measure it. Even the chemistry beneath our feet carries a new insistence: adapt, or become ornamental.

And so the Prelude stands here as a gesture of transition. It does not argue. It does not persuade. It widens. It loosens the frame through which you have been instructed to see energy, waste, sovereignty, and responsibility.

It tells you, without saying so directly, that the white paper you will read later is not a proposal but the documentation of an inevitability. And it tells you that the essay you will read next is not merely commentary but the architecture of the meaning behind that inevitability.

You will find that the essay behaves like a revelation reconstructed from familiar materials. It will ask you to consider the edges of the metropolitan map not as afterthoughts but as the places where the future incubates itself. It will show you the strange elegance of systems that no longer rely on centrality. It will reveal that the most consequential transformations now begin in the unacknowledged corners—the transfer stations, the brownfields, the sovereign plots of tribal nations, the rural corridors where feedstock accumulates with the patience of a debt the region has tried not to recognize.

Once you have moved through that narrative, the white paper—From Centralized Heat to Distributed Carbon: Repositioning Waste-to-Energy in the Twin Cities Region—will read not as a technical document but as the confirmation of the essay's prophecy. Its tables and comparisons will feel less like metrics and more like fingerprints—evidence that the logic of the future was always present, simply waiting for someone to name it.

This document is arranged deliberately:

First, the Prelude. Then, the Essay. Finally, the White Paper.

Not because that is the conventional order, but because it is the only order that allows the truth to reveal itself with dignity. Read them in sequence and you will feel the transition. Read them as one movement and you will see the structure. Read them with intent and you will recognize the inheritance. For this is not a proposal, nor an argument, nor a narrative exercise. It is a mirror held up to a region on the verge of understanding that its legacy is not ending—it is evolving. And the evolution begins here, in the quiet of this Prelude, before anything has been declared, before any data has been presented, before any argument has been made. A corridor has opened. Step inside

THE EDGE OF COHERENCE

A Strategic Treatise on Distributed Energy Sovereignty for the Twin Cities and Its Emerging Fringe

By Joe Cozart

There are cities that build their reputations on monuments, and others that build them on the silent labor of their machines. The Twin Cities belong to the latter: a pair of urban siblings that learned long ago that winter rewards the prepared and punishes the sentimental. Their thermal networks, their waste loops, their subterranean exchanges of heat and steam—none were acts of bravado. They were acts of survival disguised as engineering. And for a time, it was enough.

But infrastructure whispers its age long before its operators acknowledge it. You can hear it in the way a district pipe sighs under pressure, or in the way a combustion chamber hesitates as if reconsidering its purpose. You hear it in the public, too—those who inherited proximity without consent, whose lungs served as the unwitting interpreters of a civic argument about utility and sacrifice. The old bargain was efficient, admirable even, but bargains are hostage to context, and context ages more quickly than concrete.

What the region built was not merely an energy system—it was a narrative of control. Waste would arrive, heat would emerge, and the city would continue its choreography of winters and workdays. Yet every narrative acquires a countertext, a shadowed commentary that begins at the margins and slowly moves inward. In this case, the shadow arrived not as protest or policy, but as a simple question: What, precisely, do you intend to do with the things your legacy cannot touch?

The answer, as always, was waiting at the edge.

The SET architecture did not arise from metropolitan ambition; it arose from the quiet chaos of places where infrastructure disappears into prairie, reservation, and industrial hinterland. It is a machine that behaves like an idea: small enough to be dismissed, precise enough to unsettle, and portable enough to appear wherever contradiction becomes untenable. It does not burn so much as translate—waste into gas, gas into possibility, and possibility into an order that ignores the old gravitational pull of downtown steam and fixed assumptions.

The charm of the Twin Cities has always been its confidence that systems, once proven, will behave forever. But permanence is the most fragile of civic delusions. A district loop cannot expand into land that no longer belongs to density. An incinerator cannot comfort a neighborhood that has grown weary of abstract reassurances. And a city cannot claim carbon virtue on the basis of yesterday's chemistry.

Thus the SET platform enters not as antagonist but as revision. It does not compete with the fortress of existing thermal infrastructure; it simply refuses to be bound by the walls. It moves instead to the fault lines—transfer stations, brownfields, agricultural corridors, the liminal acreage of tribal sovereignty—those places where waste accumulates without the ceremony of civic planning. There, the machine becomes more than a device; it becomes a confession that energy, in the coming era, must learn to hunt rather than wait to be fed.

The metropolitan mind, accustomed to centrality, often misreads modularity as diminishment. Yet there is nothing small about a system that arrives fully formed, asks for so little, and answers for so much. Its syngas becomes electricity for those who cannot afford delay, heat for those who operate outside political cartography, hydrogen for an economy that has not yet decided how to describe itself. And in its quietest gesture, it offers carbon accounting sophisticated enough to embarrass its older cousins—a polite reminder that environmental virtue is no longer measured by the ton, but by the precision of one's intentions.

In the old model, infrastructure was a sermon addressed to the city center. In the new one, it is a conversation held in the margins, where the future is always less polite but far more honest. If the Twin Cities choose to listen, they will discover that their legacy was not a final act but a prelude. The pipes and boilers, the steam loops and sorting floors—they were admirable chapters, but chapters nonetheless. The story that follows belongs to the small machines that understand scale not as ambition but as relevance.

A $10 million demonstration is not a summit; it is a door. It opens onto a landscape in which counties need no longer outsource their contradictions, in which tribal lands transform waste into sovereignty, in which airports and processors and forgotten industrial corners become authors of their own energy. And once opened, doors seldom agree to be closed.

The Twin Cities built a system that warmed buildings. SET proposes a system that warms the century. One was built for a climate; the other is built for a transition. And in that distinction lies the quiet truth that every metropolis eventually confronts: the future does not arrive to flatter the past. It arrives to replace the assumptions that the past mistook for permanence.

The cities, of course, will insist they are not afraid. Cities never admit fear; they disguise it beneath task forces, feasibility studies, and administrative optimism. Yet you can sense the unease in the way their committees speak—softly, as if aware that every sentence must now perform two duties at once: to reassure the present and to negotiate with the future. There is a ritual melancholy to this, like watching a veteran conductor raise his baton over an orchestra whose instruments were tuned for a different century.

The truth is that Minneapolis and St. Paul have reached the point in their narrative when even competence becomes a liability. Legacy systems are a comfort until the moment they become an inheritance too heavy to defend. The pipes beneath St. Paul hum with the certainty of proven engineering, but certainty itself has grown expensive. The incinerator at the river's edge continues its practiced choreography of fire and steam, yet the public has learned to read smoke not as utility but as accusation. And none of these signals obey the decorum of municipal planning. They emerge from the periphery and drift inward—as most truths do.

That is where SET becomes a kind of contrarian intelligence. It does not petition the center; it circumvents it. It acts with the quiet assurance of a tool that has already accepted the verdict of the century before the century has articulated it. When a system can be placed in a brownfield, or a tribal tract, or the forgotten acreage of a county whose planners have grown tired of waiting for metropolitan benevolence, the machine becomes something more than infrastructure. It becomes an indictment of the very idea that the future requires permission.

The old guardians of the thermal order will argue for continuity, but continuity no longer behaves. Feedstock chemistry shifts like a mood, carbon expectations tighten with the velocity of a closing fist, and public patience—which once tolerated visible sacrifice in the name of invisible efficiency—no longer consents to the exchange. This is not rebellion; it is a recalibration. A system that heats a capitol dome does not necessarily heat the conscience of a neighborhood. And so the legitimacy of centralized waste-to-energy has begun to fray—not through scandal, but through irrelevance.

Gasification's elegance lies in its refusal to dramatize. It does not flare, roar, or insist on being witnessed. It operates inside its own chamber, with the discretion of a diplomatic cable written for readers who understand inference. Even its emissions behave with the modesty of a professional—cleaned before exposure, neutralized before interpretation. And because it declines spectacle, communities begin to trust it the way they trust a mechanism that keeps its intentions to itself.

This, more than any thermal efficiency or molecular advantage, is its political genius. When a system refuses to resemble the villains of a previous century—no stacks, no flames, no architectural arrogance—it disarms opposition by removing the very symbols that opposition depends upon. What remains is a machine whose presence is almost philosophical. It simply exists, performing a transformation the old order claimed was impossible without ceremony.

And that is why the Twin Cities must confront a paradox they have long postponed: their legacy systems, once proof of their pragmatism, now prove only that the future has moved elsewhere. Not out of contempt, but out of necessity. A metropolis designed around centralized combustion cannot easily explain why incineration should remain the custodian of its obligations in a decade governed by carbon arithmetic and political fragility.

The question, then, is not whether SET should enter the metropolitan frame. It is whether the metro will recognize that the frame itself has widened. Waste no longer descends into the city by default; it resides at the edges, in corridors defined not by density but by the inconvenient geography of modern life. Energy no longer radiates outward from heroic infrastructure; it accumulates in small pockets of precision where heat, power, and chemical intermediates are produced exactly where they are needed. And carbon, once the invisible cost of progress, has become the primary arbiter of whether progress can proceed at all.

What the Twin Cities once did with confidence, SET does now with strategy. It replaces scale with intention, proximity with precision, visibility with efficacy. And in doing so, it offers the region a path that is not a repudiation of its past, but a refinement of its instincts. Minneapolis and St. Paul were never wrong to build what they built. They were only unprepared for the speed at which correctness becomes obsolete.

The cities may not yet fully understand that they stand at the threshold of a calmer, more deliberate model of stewardship. But thresholds are patient. They do not demand recognition; they simply wait for the moment when the old choreography falters and a quieter rhythm takes hold. And when the shift occurs—as it always does—it will not arrive with fanfare. It will arrive with the hum of a small machine on the far edge of the metro, turning refuse into calibration, and calibration into sovereignty.

It is tempting, in moments of infrastructural uncertainty, to sentimentalize the past as if it were an argument rather than a memory. The Twin Cities, with their well-behaved blueprints and their northern insistence on competence, often fall into this habit. They recall the rise of district heating as if recalling the youth of an admired statesman—those years when pipes were laid like declarations, when combustion was the grammar of realism, and when turning waste into heat felt almost subversive in its cleverness.

Yet even the most dignified systems acquire contradictions the way old houses acquire drafts. The Hennepin facility, once heralded as a mechanical conscience, now stands in a kind of genteel embattlement—still functioning, still impressive, but surrounded by a public that no longer grants silence as a form of permission. And the district loops, those quiet arteries beneath St. Paul, pulse with a reliability that feels nostalgic precisely because the future no longer rewards reliability; it rewards adaptability.

Legacy infrastructure ages not through failure but through context. A 33-megawatt incinerator becomes, in time, a monument to what carbon used to mean. A biomass-fueled CHP plant becomes an artifact from a decade when climate was a concern rather than a condition. And so the Twin Cities now find themselves asking the most delicate of civic questions: How do you honor the virtue of what worked without mistaking virtue for permanence?

The answer begins, as all quiet reshufflings begin, at the molecular level. The difference between combustion and gasification is the difference between insistence and interpretation. One burns its feedstock as if disciplining it. The other persuades it. Heat becomes conversation rather than judgment. And within that shift lies an unexpected advantage: a technology that does not require a city to remake itself in order to remain civilized.

For SET does not demand that the Twin Cities abandon the infrastructure that once defined them. It merely reveals the unspoken truth that those systems were never designed for the world the cities now inhabit—a world in which waste is chemically mischievous, in which carbon policy behaves like an auditor rather than a philosophy, and in which public patience has become a finite natural resource.

The old model assumes a center to which waste obediently travels. The new model assumes no such etiquette. PFAS-laden residues accumulate where regulation forgot to imagine them. Plastics breed faster than policy. Agricultural byproducts move through the state like an unclaimed inheritance. And every granule of this modern reality defies the confidence of an incinerator designed in the twilight of the 20th century.

Thus SET shifts the geography of responsibility. It goes to the waste rather than summoning the waste to itself. It offers a mobility that feels almost indecent in its practicality, as if energy might finally obey the movements of the era rather than the desires of the planners. A 10-ton-per-day system suddenly matters in places that the downtown heating loops cannot reach, because the loops measure need by density, and the future measures it by vulnerability.

And in this inversion, something else occurs: a political realignment disguised as an engineering decision. For the first time, suburban transfer stations, tribal nations, rural counties, and industrial peripheries acquire access to the kind of energy sovereignty that once belonged only to the municipalities with the budget to construct their own thermal narratives. Decentralization ceases to be a slogan and becomes, instead, a form of enfranchisement.

The economics, stripped of their ceremonial spreadsheets, are equally subversive. Where the legacy systems require nine-figure commitments and thirty-year promises, the modular fleet thrives on agility. A $10 million demonstration does not mortgage a future; it investigates it. Its revenue streams—electricity, heat, synthetic fuels, tipping fees, carbon credits—behave like a portfolio rather than a pronouncement. And the payback horizon no longer competes with generational patience.

In the process, the cities discover a paradox that every aging power discovers: the elegance of the past becomes the obstacle of the present. An incinerator cannot capture carbon without re-engineering its soul. A biomass boiler cannot negotiate with evolving definitions of renewability. A steam loop cannot stretch itself into counties that no longer organize their lives around the center.

Gasification requires none of these contortions. It simply exists in the spaces where the old logic cannot travel. And in doing so, it becomes the most dangerous of technologies—the kind that makes itself inevitable not through superiority, but through relevance.

This relevance becomes clearest when examined through the lens of permitting, that peculiar American ritual in which law, science, and emotion are asked to sit together and pretend they are speaking the same language. The state's environmental gatekeepers, once content to measure compliance by parts per million, now measure it by narratives of impact and accumulations of harm. Incineration, regardless of its chemistry, has become symbol as much as facility. And symbols age faster than machines.

Gasification, with its modest casing and its absence of theatrical emissions, avoids that symbolic inheritance entirely. It slips beneath the regulatory radar not through evasion, but through appropriateness. It is the right size, the right chemistry, the right demeanor for an era in which environmental trust has become more valuable than heat itself.

Thus deployment becomes not merely possible but natural. Airports seeking decarbonization no longer require committees to imagine the improbable. Suburban transfer stations find themselves promoted from waypoints to nodes. Tribal land becomes a stage for sovereignty rather than a backdrop for state neglect. And universities—ever eager to affiliate with the aura of the future—discover in SET a laboratory that behaves like a movement.

As the system replicates across the state and region, a new architecture of energy begins to take shape—one that does not depend on the benevolence of the metropolitan core. Agricultural corridors become their own carbon accountants. Rural co-ops become their own power producers. County boards, long excluded from infrastructural destiny, become curators of their own waste-to-value symphonies.

And the cities, recognizing what has occurred, sense a truth they once preferred to postpone: the success of the edges will eventually redefine the center. Minneapolis and St. Paul will not be displaced; they will be contextualized. Their legacy systems will continue to operate, but as monuments of admirable ambition rather than arbiters of regional destiny. The new grammar of infrastructure—mobile, modular, low-carbon, chemically fluent—will dictate the tempo.

The demonstration facility thus becomes less an experiment than a declaration: that the future of energy belongs to those who understand that permanence is merely a stage between iterations. And what the Twin Cities once did for thermal ingenuity, they now have the opportunity to do for distributed carbon performance—

lead, not because they must, but because history has always rewarded those who accept its momentum rather than resist it.

There comes a moment in every mature infrastructure system when it must confront the truth that it has become too large to move and too historic to innovate. The Twin Cities reached that moment not through failure, but through the gravity of their own competence. Their centralized facilities were built so well, so confidently, that they could not imagine a world in which their very scale would become their most delicate vulnerability.

The incinerator, aging with the authority of a once-favored statesman, persists through compliance but not conviction. It meets the letter of the permit, but the permit itself belongs to an era that has dissolved beneath the weight of its own outdated assumptions. Public patience, once infinite for anything that kept the lights warm through winter, now studies every plume, every particulate, every whiff of legacy infrastructure with the scrutiny once reserved for scandal.

And so the great paradox manifests: what was once regarded as the height of urban environmental stewardship is now interpreted through a different political lens—one that measures not efficiency, but inheritance. A facility that processes 50% of the county's waste may remain essential, yet its essentiality has become controversial simply because it is visible. Visibility, in the 2020s, is no longer a marker of trust; it is a liability.

Gasification, by contrast, understands the new etiquette. It does not demand to be seen. It does not ask the public to tolerate a skyline gesture. It does not rely on the theater of steam. It opts instead for discretion. The machine resembles nothing more provocative than a set of containers arranged in a configuration only engineers fully appreciate. And because it does not resemble the ghosts of an older energy order, its presence is permitted to exist without the emotional surcharge that accompanies older thermal institutions.

This refusal to dramatize becomes the key that unlocks every door the old systems can no longer open. Permitting authorities, fatigued by political combat around incineration, discover in SET a technology that requires no philosophical apology. Environmental justice advocates find themselves confronting a system that generates no visible betrayal. Lawmakers, eager to claim progress without igniting resistance, introduce it gently into their vocabulary. Even counties that once recoiled at the idea of hosting thermal equipment begin to reconsider, now that the equipment looks more like infrastructure and less like industrial confession.

But technology alone does not reshape a region; economics do. And here, too, SET reveals the quiet subversion embedded in its logic. Legacy systems were born of an era when infrastructure was synonymous with permanence—when political capital, financial capital, and moral capital converged into monumental projects designed to last lifetimes. The amortization schedules were measured in decades, the returns in civic pride.

Today, permanence reads as rigidity, and rigidity as risk. Investors no longer admire the grandeur of a multi-hundred-million-dollar plant when a mid-eight-figure fleet can accomplish more, faster, with less political exposure. Municipalities no longer crave singular assets when modular ones provide resilience without demanding fealty. Even utilities, long conditioned to think in generational increments, are discovering that the path to decarbonization is not vertical but lateral—not concentrated but distributed.

Gasification introduces an economic logic that is unnervingly contemporary: start small, scale horizontally, relocate if needed, refine continuously, monetize carbon, translate waste into fuels that behave like currency, and allow each site to fail or flourish without endangering the integrity of the whole. It is the infrastructure equivalent of a diversified portfolio—steady, flexible, immune to catastrophes of singular dependency.

In such a world, the old argument for centralized plants begins to look not wrong, but incomplete. The cities built machines for an era that believed stability was the highest virtue. The century that followed taught them that adaptability is the higher one.

And so the region moves, almost reluctantly, toward a future arranged by micro-hubs rather than monuments. Transfer stations once seen as logistical afterthoughts become points of origin for distributed clean fuels. Brownfields that once symbolized industrial injury become laboratories of chemical reclamation. Tribal land, long treated as peripheral, becomes central to the logic of energy sovereignty. Airports, universities, industrial zones—each becomes capable of hosting a system that asks for so little and answers for so much.

The modular network begins to resemble not infrastructure, but strategy.

And as with any strategy worth remembering, replication becomes its instrument. For once the Twin Cities accept that a system can be both small and consequential, the state will discover that such systems require no ceremony to proliferate. Like a quiet idea that moves from conversation to conversation, SET spreads through the rural corridors, river towns, agricultural economies, tribal sovereignties, and post-industrial geographies that have long been waiting for a technology that does not ask them to mimic the urban core to claim the future.

And in this replication, a new balance of power emerges—one the metro did not expect, but may ultimately embrace. The energy map of Minnesota begins to invert. The influence of the center does not disappear; it becomes one node among many, rather than the gravitational axis of all. Waste, once treated as a burden to be centralized, becomes a currency to be localized. Carbon, once an afterthought of combustion, becomes the metric of intention.

In this reordering, the cities stand not as relics but as prologue. Their legacy infrastructure becomes a monument to the ingenuity of an earlier chapter—admirable, dignified, and insufficient. SET becomes the chapter that follows, not because it seeks dominion, but because it seeks relevance. And relevance, in this century, has become the most powerful currency of all.

The cities, if they were honest, would admit that they have always been ambivalent about the edges. The center made sense to them—dense, visible, accumulating political gravity with every building stacked upon the last. But the edges, those mutable stretches of counties and cooperatives and sovereign lands, refused to be organized by metropolitan logic. They generated waste with enthusiasm but received little infrastructure in return. They carried essential industries but were not treated as essential. They bore contamination without receiving ownership of the solutions. In short, they were the parts of the map the city preferred to interpret rather than understand.

This imbalance endured for decades because the old thermal systems required it. Centralization imposes hierarchy. It demands that power and waste perform pilgrimage to a designated altar. It rewards density and punishes dispersion. And so the edges remained places of perpetual waiting—waiting for budget, waiting for heat, waiting for investment, waiting for the neon benevolence of an urban government always slightly too preoccupied to remember that the cold bites just as hard beyond the city limits.

SET disrupts this not through revolution but through indifference. The machine does not care where it is placed. It does not require density to justify its existence. A parking lot with a suitable concrete pad will do. A decommissioned industrial site will do. A rural transfer station, ignored for twenty years, will do. A corner of tribal land that the county pretended was peripheral will do. In this way, the technology behaves with a kind of democratic precision. It distributes capability to places that had long been expected to supply problems but not solutions.

This redistribution terrifies the old order in a way it cannot articulate. If energy sovereignty can emerge from a ten-ton-per-day deployment outside the metropolitan frame, what becomes of the center's traditional authority? If rural counties no longer need the city to digest their waste or validate their carbon practices, how does the city maintain its imagined role as the arbiter of modernity? And if tribal nations, long treated as exceptions to the rule, become early adopters of chemical conversion technologies that outperform incineration on emissions, economics, and politics, then the rule itself collapses.

The old model depended on the periphery's dependence. The new model eliminates dependence altogether. It replaces the center-periphery relationship with a constellation—small, self-contained, sovereign, and quietly potent.

This constellation strengthens the entire region. A single distributed system reduces trucking miles, lowering both cost and carbon. A cluster of systems creates redundancy that legacy infrastructure can only simulate through expense. And the more installations emerge—at airports, on university grounds, in agricultural corridors—the more the state begins to resemble a network rather than a hierarchy.

Nowhere is this more evident than in the chemistry that governs the 21st-century definition of waste. PFAS, plastics, volatile blends, agricultural residues—each one a reminder that modernity produces materials that refuse to obey the civic optimism of the 1980s. Incineration was designed for an era when the word "waste" implied disposability. Gasification was designed for an era when the word "waste" implies danger, opportunity, or both.

When cities confront PFAS, they debate destruction temperatures, transport liabilities, and legal exposure. When gasification confronts PFAS, it simply places the molecules in an environment engineered for their disassembly. Where the incinerator must prove it can handle the exception, the gasifier was built for the exception from the start.

This is why regulators, once skeptical of anything involving heat, now examine SET with unusual neutrality. The system enters their offices not as an applicant but as an answer—one that aligns with federal incentives, university research agendas, tribal sovereignty statutes, brownfield remediation programs, and the quiet desperation of counties tired of shipping their contradictions to a city that no longer wants them.

Thus the technology gains allies without courting them. The agricultural co-ops see dryer fuel. The rural counties see avoided landfill fees. The tribal governments see sovereignty. The universities see grants. The airports see decarbonization optics. The utilities see distributed peaker displacement. And the metro, which once imagined itself the protagonist of all regional narratives, begins to sense that the supporting cast has grown unexpectedly powerful.

What remains, then, is the final, unspoken truth—the one the white paper never says explicitly but implies with every comparison, every table of economics, every line of strategic alignment. The Twin Cities are not being replaced. They are being relieved. Unburdened. Freed from the expectation that they must solve the waste and energy problems of a geography far larger than their political imagination.

When the $10 million demonstration arrives—whether in the metro or just beyond its reach—the city will feel something it has not felt in decades: the sensation of an obligation lifted. The sensation of no longer being the sole custodian of the region's contradictions. And beneath that sensation, an even deeper one: the recognition that the future is not a threat but a transition.

The demonstration will turn waste into fuel with an elegance the old systems cannot perform. It will turn skepticism into policy, policy into capital, capital into replication, replication into normalcy. And as the machine hums quietly at the edge—transforming its feedstock with the discreet confidence of something that knows it has arrived at precisely the right moment—the center will understand that this is not a rebellion. It is an inheritance.

The Twin Cities built the first chapter. SET writes the next because the century demands it.

What the region will soon discover is that alignment never begins with consensus; it begins with necessity. And necessity, once it gathers enough quiet momentum, becomes indistinguishable from inevitability. The actors who believe they are choosing to participate in the next phase of the state's energy evolution will eventually realize that they were chosen—not by politics, but by physics, economics, and the slow uncoiling of public patience.

The counties, those steady custodians of the mundane burdens of modern life, will come first. They have always lived closest to the problem—hauling fees rising, landfill space diminishing, rural roads shuddering under the weight of cities exporting their decisions to the outer ring. When offered the chance to convert waste not into a political grievance but into a revenue stream, they will not hesitate. They will see in SET something they have not seen in decades: a future that does not require them to wait for the city's permission.

Then come the metropolitan planners, those architects of long-range strategy who speak in the polite cadences of feasibility and resilience. They will arrive slowly, at first convinced that only megastructures carry legitimacy. But as the small systems begin to accumulate—one on tribal land, one adjacent to a co-op, one quietly operating at an airport—they will recognize that legitimacy now accrues to humility. And nothing is more humble, or more strategically dangerous, than a machine that solves a problem without demanding a parade.

The universities will follow in their perennial way, attributing to themselves the foresight they never possessed. They will produce studies, convene panels, and assign graduate students to measure precisely what the machine is already doing. They will call this research; the machine will call it observation. And yet the collaboration will benefit all parties. Nothing accelerates acceptance like academic proximity.

Then come the utilities—pragmatic creatures whose public rhetoric leans toward vision, while their private instincts lean toward reliability. They will not embrace SET because it is fashionable; they will embrace it because it solves the problem they most fear: how to maintain authority in a world that no longer tolerates monolithic dependence. Distributed peaker displacement, hydrogen blend studies, modular microgrids—these will become the polite euphemisms through which utilities admit that the future can no longer be centralized.

And above them all, sovereign nations—tribal governments that have waited with dignified restraint for technologies that respect both their land and their autonomy. SET will be the first system that feels as if it were designed with sovereignty in mind. The machine requires no subordination to metropolitan hierarchy, no dependence on state permission, no capitulation to outdated energy norms. It simply offers capability—clean, compact, and quiet—and sovereignty recognizes itself in that gesture.

One by one, these actors form a constellation, and the constellation forms a quorum. And once a quorum forms, the center is no longer the center; it becomes one star among many, each contributing light but no longer monopolizing direction.

What emerges is not merely an infrastructure model, but a philosophy. It is the slow realization that the future of energy, waste, and carbon is not a question of scale but of orientation. That systems must move toward the problem, not demand that the problem come to them. That sovereignty is not symbolic but chemical. That carbon is not a pollutant but a metric of intention. And that power—real power—belongs not to the structures that rise above the landscape, but to the ones that embed themselves quietly within it.

When the demonstration facility finally activates—somewhere on the metro edge where the old cartography loses confidence—the cities will feel a shift. Not a rupture, not a revolution, but a subtle correction, like a compass needle acknowledging a stronger magnetic truth. The machine will convert its first batches of waste into clean syngas, then into electricity or heat or hydrogen or methanol, depending on who asks. And in that transformation, the region will glimpse the future not as aspiration, but as practice.

From that point forward, the question will not be whether SET should scale, but how fast.

And the legacy systems—the incinerator, the biomass boiler, the district loops—will continue their dignified service, no longer as monuments to an era of centralized control, but as honored elders of a lineage that has learned to evolve.

The cities will not mourn this transition. They will find in it a quiet relief, the kind one feels when a long-held responsibility finally finds a more agile steward. And the region will discover that the future was never waiting in the city at all. It was waiting at the edge—impatient, precise, and ready.

The moment the first deployment succeeds—and it will succeed, not because optimism demands it but because physics does—the region will enter a new phase of its identity. Success, after all, is the most persuasive form of politics. Once a machine on the edge proves that waste can be reinterpreted into energy, that carbon can be measured rather than excused, that sovereignty can be exercised rather than requested, the center will feel an unmistakable shift in the air. A shift that cannot be influenced, regulated, or discouraged. A shift that behaves like an arriving truth.

At first, the cities may try to label it. Bureaucracies have always believed, somewhat romantically, that naming a thing grants dominion over it. They will call it a pilot, then a demonstration, then a supplemental resource, and eventually a "non-oppositional parallel infrastructure," as if language alone could bind a system designed to evade the gravitational pull of centralization. But the machine does not respond to naming. It responds only to need.

And the need, embarrassingly for those who still cling to the monuments of the past, is not in the places where the monuments stand. It is in the industrial outskirts where feedstocks pile without ceremony. It is in the rural counties where infrastructure budgets collapse under the weight of obligations that cities exported long ago. It is in the agricultural processors who have grown weary of pretending that propane is destiny. It is in the tribal nations who understand that sovereignty, to be meaningful, must extend to the molecular level. It is in the airports whose runways stretch into a century that no longer tolerates visible contradiction.

Each site discovers, almost accidentally, that SET is not an intervention but a conversation—a negotiation between what the region once believed about waste and what the region must now believe about carbon. The machine does not lecture; it interprets. It does not impose; it adapts. It does not declare; it proves.

And once it has proven itself in enough corners of the map, the demonstration ceases to be a demonstration. It becomes a pattern. And patterns, once recognized, acquire a power that no centralized institution can outmaneuver. They reveal that the future is not being constructed at the center but is quietly accumulating at the boundaries, where innovation is neither ceremonial nor ideological, but simply necessary.

It is here that the cities find an unexpected clarity. The incinerator is not a failure; it is a testament. The district loops are not obsolete; they are honored. But their authority ends at the limit of their geography, and the era has changed its geography altogether. The new landscape is scattered, irregular, and opportunistic—defined not by density but by friction. SET thrives in friction. It metabolizes it.

And thus a new regional order emerges, one that resembles a quietly humming constellation stretching from the metro outward in all directions. Each node carries its own chemistry, its own economics, its own sovereignty. Each one contributes to a carbon topology that no longer depends on a single stack, a single grid-tie, a single thermal loop. And each one, by existing, diminishes the emotional dependency the cities once demanded from the rest of the state.

In this distributed order, waste stops behaving like an obligation and begins behaving like leverage. Carbon stops behaving like a liability and becomes a ledger of intention. Energy stops behaving like a centralized privilege and becomes a local right. And sovereignty stops behaving like a ceremonial phrase and becomes the quiet posture of a region that has discovered how to govern itself at the molecular level.

The $10 million demonstration becomes the hinge upon which this transformation turns. It is neither large enough to frighten nor small enough to dismiss. It is the ideal size for a century that mistrusts grandeur but respects decisiveness. It shows the cities how to relinquish without losing stature. It shows the counties how to assume responsibility without inheriting burden. It shows the tribes how to convert autonomy into industry. And it shows investors that resilience is no longer a poetic ambition but a measurable commodity.

When the system replicates across Minnesota and into the Upper Midwest, the region will finally understand the point the white paper could only imply: the future was never a question of replacing the old model. It was a question of completing it. Flanking it. Liberating it from the expectation that it must be everything, everywhere, for everyone.

The Twin Cities built a system that defined an era. SET builds a system that defines the one that follows. And the quiet genius of the transition is that both systems will coexist—one as the architectural memory of what the region once needed, the other as the architectural instinct of what it needs now.

When the machine hums at the edge of the metro, turning discarded material into distributed carbon precision, the region will feel, perhaps for the first time, that the future is no longer approaching.

It has arrived.

The arrival of the future is seldom theatrical. It does not descend in a blaze of announcements or assemble itself in front of cameras. More often, it appears at the periphery disguised as something modest—a pilot, a prototype, a containerized system humming in a corner of land no one had bothered to politicize. Yet modesty has always been the most reliable courier of change; it obscures the transition long enough for the transition to become irreversible.

And so the Twin Cities will find, in time, that what they mistook for a discreet experiment was in fact a redefinition of their role. Not a diminishment—nothing so dramatic—but a rebalancing. A recognition that the privilege of centrality was an artifact of a century that depended on it. A whisper that the new century does not bow to the same gravitational pull.

The machine at the edge—quiet, measured, strangely elegant in its industrial humility—will not ask the cities to surrender their identity. It will simply reveal that identity to be smaller than the geography it once commanded, and larger than the boundaries it once imagined. For in a distributed order, the center is not obsolete; it is contextual. A participant among peers rather than an axis surrounded by dependents.

This shift in posture—subtle yet decisive—will be the true legacy of the demonstration. Not hydrogen output or carbon intensity metrics or avoided landfill tonnage, though all will be impressive. The lasting legacy will be the cities' newfound ability to act without burden. To collaborate without domination. To influence without requiring obedience. The incinerator, the biomass plant, the steam loops—they will remain pillars, but pillars among many, not the lone custodians of a region's energy narrative.

Around them, the constellation of micro-hubs will expand. A tribal installation treating contaminated waste on sovereign land under sovereign logic. A rural deployment processing agricultural residues in a county that once exported its contradictions. A university site converting policy theory into chemical reality. An airport turning operational waste into a demonstration of decarbonization. A suburban transfer station discovering it need not ship its burdens elsewhere to negotiate its responsibilities.

Each node will hum with the kind of quiet authority that does not need applause to justify itself. And because none of them depend on the center, the system as a whole will become something the old model could never be: self-correcting. Redundant by design. Resilient not by ambition, but by arrangement.

The public will feel the difference first—not as a spectacle, but as an absence. Fewer debates about incineration. Fewer nighttime flares. Fewer truck convoys threading the city with the choreography of obligation. The energy conversation will shift tone, abandoning the melodrama of crisis for the calm precision of stewardship. Citizens, no longer forced to choose between denial and despair, will discover that the architecture of the future is neither heroic nor apocalyptic. It is simply appropriate.

Investors, ever attuned to the scent of inevitability, will move next. They will see in distributed conversion the rare combination of low political friction, high replicability, and steady, multi-vector returns. In a world increasingly allergic to megaprojects, a $10 million deployment becomes not an investment but an opportunity to anchor a portfolio in relevance rather than nostalgia. Capital loves systems that scale horizontally; SET was built for that affection.

Policymakers will follow, though they will pretend they were leading. They will craft frameworks that appear to guide the technology but in truth only describe what the technology has already made possible. They will write legislation that reads like prophecy but is merely the transcription of what they observed too late to influence. This is not an indictment; it is the natural choreography of governance in an era where innovation has outrun bureaucracy.

And when the constellation becomes a network—and the network becomes a norm—the cities will look back on their old systems not with regret, but with gratitude. For without those systems, the region would never have learned to think thermally, chemically, infrastructurally. Those legacies were the apprenticeship. SET is the vocation.

The story, then, is not one of replacement but of succession. The Twin Cities demonstrated what intelligent infrastructure looks like. SET demonstrates what adaptive infrastructure feels like. And the region, in accepting both, becomes something rare in American civic life: a place that remembers its past without condemning itself to live in it.

The demonstration will not be the beginning. It will not be the end. It will be the hinge—a moment when the future stops tapping politely at the door and steps inside, unannounced, carrying with it the quiet authority of something that no longer requires introduction.

And from that moment forward, the region will operate under a new, unspoken assumption: that carbon, waste, sovereignty, and resilience are no longer problems to be solved at scale, but truths to be interpreted at the size of the communities that live them.

The Twin Cities once built the backbone of an era. SET builds the nerve system of the next. And the two, together, form a body of infrastructure finally equal to the century that awaits it.

In the end, it is not technology that changes a region; it is recognition. The moment a community realizes that the future is no longer something to prepare for but something that has already begun to occur around it, the transformation is complete. The Twin Cities will reach that moment gently, without ceremony, without the grand gestures their older infrastructure once required. It will arrive the way winter arrives here—quietly, steadily, and with an authority that needs no explanation.

They will see a machine operating on the periphery with an ease that borders on indifference, turning the overlooked into the essential. They will see waste—once transported, sorted, burned, justified—suddenly reinterpreted into a fuel whose cleanliness feels almost rhetorical. They will see carbon—once a hidden arithmetic—displayed in the open as a ledger of choices made deliberately rather than accidentally. And they will see that the institutions they once expected to lead the next chapter have already begun, subtly, to defer to the elegance of the smaller, the faster, the more precise.

For this is the quiet truth that the century has imposed upon all who build: progress no longer rewards scale. It rewards relevance. The structures that endure are not the largest, but the ones most willing to adapt without demanding applause. And SET, with its modest casing and immodest implications, is nothing if not an embodiment of this new discipline.

What the region gains is not merely a technology. It gains a posture. A way of thinking that does not demand loyalty to the past nor disloyalty toward it, but simply acknowledges that the past's brilliance was never meant to freeze the future in place. The cities will keep their loops and their boilers, for they are still valuable and still honorable. But they will no longer pretend that those systems alone can carry a century defined by irregularity, imbalance, and chemical truth.

The counties will gain autonomy. The tribes will gain sovereignty in practice rather than ceremony. The rural processors will gain heat and power without petition. The airports will gain a decarbonization tool that does not require political choreography. And the investors will gain something rare in the landscape of modern infrastructure: a system that scales by multiplication rather than justification.

This is the quiet brilliance of the $10 million demonstration. It is not expensive enough to frighten anyone, nor trivial enough to ignore. It occupies the exact center of American plausibility. It proves the possible without requiring the mythical. And once proven, it refuses to return to the realm of the theoretical.

The Twin Cities will not be asked to choose between the legacy they built and the future that has arrived. The choice has already been made by a world that no longer tolerates slow carbon, rigid infrastructure, or the idea that one system can serve all geographies equally. The cities need only acknowledge that the baton is passing—not from competence to novelty, but from singularity to constellation.

And so the century turns. Not dramatically. Not violently. But with a precision that feels almost like grace. A machine hums at the metro's edge. A county recalibrates its waste. A tribe confirms its sovereignty. A university publishes a study. A utility amends its forecast. A planner redraws a boundary. A city exhales.

And the region, at last, becomes what it was always capable of becoming: a place where infrastructure is not a monument but a mirror—reflecting, with uncommon clarity, what a people choose to value when the world no longer permits hesitation.

This is the work ahead.

This is the invitation.

This is the moment to begin.

ONE-PAGE BUSINESS CASE FOR COUNTY BOARDS

Every county faces a moment when the arithmetic of the past stops corresponding to the burdens of the present. Hauling contracts grow heavier. PFAS liabilities accrue silently beneath the ledger lines. Landfill lives shorten year by year. And residents — with increasing impatience — expect their local government to solve problems created by distances and decisions beyond their control.

SET Smart Gasification enters that moment with uncommon precision. It does not ask the county to build a megastructure. It does not require a referendum or a reconfiguration of the tax base. It requires space, feedstock, and a willingness to interpret waste as a local resource rather than an exported liability. Its ten-ton and twenty-five-ton deployments match the scale of county operations rather than the ambitions of the urban core.

For county boards, the business case resolves into a simple clarity: the cost of continuing the old model is now greater than the cost of establishing the new one. SET reduces hauling miles, stabilizes operating expenditures, destroys PFAS at the molecular level, and aligns with state and federal incentives designed explicitly for distributed decarbonization. The county that adopts it first will not merely solve a waste problem — it will gain an asset that reshapes the economic and political logic of its jurisdiction. This page exists to express that reality plainly, without ornament: the counties at the edge are no longer passive recipients of metropolitan decisions. They are now the rightful authors of their own infrastructure. SET makes that authorship possible.

What the narrative has opened, the evidence now confirms.

What was atmosphere and interpretation now becomes architecture and proof.

The white paper begins.

GMJoe™ Consulting
Joe Cozart | Clarity Consultant
Email: joe@gmjoe.org | Mobile: 701-885-2684
Website: www.gmjoe.org

GMJoe™ Consulting

Author's Note on Scope and Use

This work presents an analytical and conceptual exploration of system design, energy conversion, and applied infrastructure strategy. It is intended to provide clarity, structure, and perspective on emerging approaches, including SET smart gasification.

It is not a technical manual, engineering specification, or site-specific recommendation.

Any systems, technologies, or frameworks discussed are presented for informational purposes only. No representations or guarantees are made regarding performance, outcomes, or suitability for any particular application.

Nothing in this work should be interpreted as professional, engineering, environmental, or financial advice. Implementation of any system described requires independent evaluation, site-specific analysis, and appropriate technical and regulatory consultation.

References to technologies or systems developed by others do not imply authorship, endorsement, or validation by those parties unless explicitly stated.

White Paper:

"From Centralized Heat to Distributed Carbon: Repositioning Waste-to-Energy in the Twin Cities Region"

by GMJoe™ Consulting

1. Executive Introduction

Every city tells a story through its infrastructure. In Minneapolis and St. Paul, the story has long been one of midwestern pragmatism, quietly elegant engineering, and a civic willingness to do what makes sense—long before it becomes fashionable. District heating in the Twin Cities is not an experiment; it is a fully operational system, built on decades of investment in combined heat and power (CHP) infrastructure. It powers hospitals, schools, office towers, and city government buildings. It recycles heat, reduces landfill volume, and has made

Minneapolis one of the few American cities with a coherent urban thermal strategy.

But coherence is not permanence. The conditions that once made centralized CHP the rational choice are shifting. Waste composition is more volatile. Carbon policy is tightening. Public perception has evolved. And critically, the geography of opportunity is moving—away from city centers, and out toward the edges of the grid, the rural waste streams, the remediating soils, and the unpowered margins where infrastructure has not yet caught up to ambition.

This white paper, prepared by GMJoe™ Consulting, explores the case for a next-generation complement to the Twin Cities' centralized systems: the SET Smart Gasification Platform. Modular, mobile, and carbon-precise, SET is not a competitor to Hennepin County's Energy Recovery Center (HERC) or St. Paul's district heating loops. It is a parallel system, designed not for where the grid is, but for where energy, remediation, and resilience must go next.

The timing is not accidental. As Minneapolis grapples with rising scrutiny of incineration emissions, as the MPCA tightens its permitting lens, and as tribal, agricultural, and suburban communities seek solutions outside of capital-bound infrastructure, a new model is needed. One that can take what we discard—not just urban waste, but policy blind spots, contaminated feedstocks, and logistical inefficiencies—and convert it into local power, clean fuel, and carbon performance.

We will trace the evolution of Minneapolis-area CHP infrastructure, unpack the design and advantages of the SET system, and offer a side-by-side comparison rooted not in ideology, but in deployment logic, fiscal prudence, and regulatory reality. The analysis culminates in a proposal: that a $10 million demonstration of SET's smart gasification system in the Twin Cities metro—or at its energy edge—could anchor a new era of distributed waste valorization and rural-industrial carbon stewardship.

The legacy of HERC and the St. Paul heating network is one of infrastructure done well. This paper argues that SET represents the next chapter: infrastructure built for what comes next.

2. Background and Context

To understand the path forward, we must first understand the elegance of what already exists. Minneapolis and St. Paul are among the few American metropolitan regions to have fully embraced the concept of urban heat as infrastructure. Beginning in the 1980s, planners and engineers in the Twin Cities recognized that waste—already being hauled, sorted, and incinerated—could do more than disappear. It could heat buildings. It could stabilize electrical supply. It could form the thermal backbone of a city built for winters.

The flagship of this approach is the Hennepin Energy Recovery Center (HERC), a 1,212-ton-per-day waste-to-energy facility located in downtown Minneapolis. HERC generates roughly 33 megawatts of electricity—enough to power 25,000 homes—and provides steam to the downtown heating loop, reducing demand on fossil gas during peak months. On the other side of the river, District Energy St. Paul uses a CHP plant, biomass boiler, and solar thermal array to deliver heating and cooling to over 200 buildings, including the State Capitol.

These systems are not theoretical. They are embedded. HERC alone processes nearly half of Hennepin County's municipal solid waste. The St. Paul system is one of the largest urban biomass-fueled networks in North America. Together, they reflect a legacy of public-private foresight: local government, utilities, and engineering firms creating long-term value from materials that most cities continue to bury.

But legacies are not immune to friction. HERC has faced increasing community opposition, particularly from North Minneapolis residents concerned about air quality and environmental justice. The facility's 30-year-old emissions systems, while still compliant, are rooted in a regulatory era that predates modern climate expectations and cumulative impact modeling. Meanwhile, district energy expansion is constrained by the very success of the original network: fixed infrastructure, aging pipes, and an economic model that assumes population density where none is growing.

At the same time, new challenges are surfacing:

- Feedstock complexity: The waste stream is evolving—lighter packaging, more plastics, more contamination.

- Policy pressure: The Minnesota Pollution Control Agency (MPCA) is under pressure to prioritize lower-emission, lower-impact technologies in environmental justice zones.

- Carbon targets: The state has committed to 100% clean electricity by 2040—requiring every megawatt to justify its carbon intensity and lifecycle.

What was once efficient may now be insufficient.

Into this context steps the SET Smart Gasification platform—not as a critique of Minneapolis's approach, but as a tool for the next phase. A system built not for the dense grid alone, but for the interstitial spaces where waste still accumulates, power is still needed, and legacy infrastructure cannot reach.

This is the opportunity: to transition from a central model of thermal control to a distributed model of carbon performance. And to do so not through wholesale replacement, but through strategic augmentation—bringing modularity to a system built on fixed expectations.

3. The SET Smart Gasification Platform

There is a temptation in legacy-rich cities to treat new technology as a risk. But what if the real risk is assuming the old systems will always apply?

The SET Smart Gasification platform does not ask Minneapolis to abandon its thermal identity. It asks: what about the waste that HERC can't take? What about the heat that district loops can't reach? What about the carbon we now need to track, report, and monetize—not just combust and scrub?

SET's platform was built not to serve the center of the city, but the periphery of the system. Its architecture is modular, containerized, and scalable, ranging from 1 to

50+ tons per day. It does not require a grid—it creates one. It does not require a district—it supplies one.

At the core of the system is a high-temperature gasifier that converts solid and semi-solid waste into a synthetic gas composed primarily of hydrogen, carbon monoxide, and methane. Unlike incineration, which combusts waste in oxygen-rich conditions to generate heat, gasification operates in an oxygen-limited environment. The result is not just energy—it's chemical feedstock, clean combustion potential, and carbon accounting leverage.

Key features include:

- Feedstock flexibility: SET accepts MSW, agricultural residues, treated wood, biosolids, plastics, and PFAS-contaminated material. No front-end homogenization or RDF processing is required.

- Output optionality: Syngas can be used for on-site power (via engines or turbines), hydrogen purification, methanol production, or CHP integration. Output is adaptable to market demand.

- Emissions control: Pollutants are captured in the syngas stream before combustion. This pre-combustion cleaning approach dramatically reduces stack emissions and enables easier compliance in EJ-designated areas.

- Deployment speed: Units can be sited, permitted, and commissioned in under 12 months—especially on industrial or brownfield land.

- Digital integration: Real-time monitoring, remote diagnostics, and AI-supported operations allow fleetwide performance management.

Unlike HERC, which relies on a centralized model and fixed-scale economics, SET platforms are designed to move toward waste, not wait for waste to arrive. That mobility means resilience. It means remediation. And it means policy alignment with state and federal priorities pushing for distributed, low-carbon, renewables-compatible systems.

In a Twin Cities context, SET can:

- Serve underserved or waste-stranded zones.
- Convert problematic feedstocks into clean local energy.
- Integrate with hydrogen hubs and biofuel corridors.
- Support the transition to carbon-labeled municipal services.

SET is not the next incinerator. It is the next assumption: that waste, wherever it exists, can be transformed into local carbon precision.

4. Comparative Technology Analysis

When evaluating energy infrastructure, the temptation is to compare what is visible: stacks, megawatts, thermal loops. But what matters now is what's invisible—carbon intensity, feedstock resilience, modular risk profiles, and the ability to flex with policy, weather, and market demand.

Let us begin with the known.

Hennepin Energy Recovery Center (HERC)

- Design: Mass-burn incineration
- Feedstock: Sorted municipal solid waste
- Output: ~33 MW electricity + steam to district heating
- Emissions: Post-combustion control; meets permit, but dated
- Fixed Infrastructure: Centralized location, long-haul waste intake
- Public Perception: Increasing scrutiny in EJ zones
- CapEx: Fully amortized, but no feasible expansion without overhaul
- Carbon Role: No current carbon capture; limited future flexibility

District Energy St. Paul (CHP)

- Design: Biomass boiler + CHP + solar thermal
- Feedstock: Wood chips, some refuse-derived fuel

- Output: Heat and cooling for 200+ buildings
- Emissions: Well-managed, but biomass under increasing carbon review
- Infrastructure: Centralized thermal loop, pipe network dependency
- Flexibility: Limited to adjacent zone
- CapEx: High sunk cost; expansion slow
- Carbon Role: No CCS; limited fuel flexibility

Now, contrast this with:

SET Smart Gasification

- Design: Modular high-temperature gasification
- Feedstock: Mixed waste, plastics, biomass, contaminated soils
- Output: Syngas for electricity, hydrogen, methanol, or process heat
- Emissions: Pre-combustion gas cleaning; stack optional
- Infrastructure: Portable, site-scalable, containerized
- Public Perception: No incinerator stigma; potential climate-positive

- CapEx: ~$10M for full system; scales by module
- Carbon Role: Native compatibility with carbon capture, LCFS, 45

Key Differentiators

Attribute	HERC / District Energy	SET Smart Gasification
Deployment Model	Centralized, fixed	Distributed, modular
Feedstock Range	MSW, biomass	PFAS waste, plastics, biomass, MSW
Carbon Adaptability	Retrofit-dependent	Native CCU, hydrogen-ready
Public Viability	Mixed, opposition zones	Favorable in EJ and brownfield sites
Energy Flexibility	Heat + electricity	Syngas to electricity, H_2, methanol
Deployment Speed	Multi-year	<12 months
Policy Alignment	Legacy fit	Aligned with 2020s–2035 targets

The result is not a condemnation of Minneapolis's centralized infrastructure. It is a clarification: legacy systems are bounded by their era, and their strengths—thermal efficiency, density leverage, operational history—do not scale to new geographies or policy environments. SET, by contrast, was born of constraint. Its architecture is inherently adaptive, built not to replace the grid, but to accompany it into the future.

5. Economic Model and Capital Requirements

Infrastructure is a declaration of priorities—made in concrete, steel, and amortization schedules. For Minneapolis and St. Paul, the investment in centralized CHP was rational, disciplined, and effective for its time. But that model—capital-heavy, site-bound, and built for 30-year recovery—now collides with a world that requires fast carbon pivots, smaller deployment footprints, and multi-vector revenue flows.

Minneapolis CHP Economics (HERC, St. Paul District Energy)

- Estimated CapEx:
- HERC: Originally ~$160M (1989 USD, unadjusted); expansion infeasible without >$300M reinvestment.
- St. Paul CHP: Similar scale; biomass retrofit in 2000s cost ~$35M.
- Revenue Model:
- Power purchase agreements with Xcel Energy.
- Steam sales to public and commercial buildings.
- Tipping fees from waste haulers.

- Cost Recovery:
- 25–30 year amortization.
- High sunk cost dependency.
- Constraints:
- Expansion blocked by pipe geography, urban land use, and rising opposition.
- Any new investment must meet 2024 permitting hurdles and 2040 carbon rules.

SET Smart Gasification Economics

- Demonstration Facility CapEx (~10 TPD):
- $4.5M – Gasifier units (2–3 modules)
- $2.0M – Syngas cleaning, energy conversion (genset, turbine, fuel cell)
- $1.5M – Site prep, utilities, civil works
- $500K – Monitoring, remote telemetry, safety systems
- $500K – Grid integration or fuel purification (H_2 or MeOH)
- $1.0M – Contingency and commissioning
- Total: ~$10M
- Revenue Streams:

- Electricity: ~$0.09–0.11/kWh (local co-ops or behind-the-meter agreements)
- Heat: Direct sales to adjacent partners (dryers, ag processors, industrial sites)
- Tipping Fees: $50–$95/ton (rural or niche waste sources)
- Carbon Credits: 45Q, LCFS, IRA hydrogen and renewable fuel incentives
- Synthetic Fuels: Methanol at $400–600/ton; H_2 at $3–$6/kg (depending on offtake partner)
- ROI Profile:
- 5–7 year payback depending on fuel pathway
- Scalable modules reduce upfront risk
- Portability supports asset redeployment or multi-site use

Where Minneapolis-area CHP requires long political and financial commitment to a single site, SET enables capital agility: investors, municipalities, or federal partners can enter at low volume, validate performance, and expand horizontally, not vertically. This enables risk-staged investment, targeted deployment, and adaptive return models—far better suited to post-industrial geography and fast-moving policy windows.

In short: the Minneapolis system is infrastructure as fortress. SET is infrastructure as fleet.

6. Policy and Permitting Landscape

In infrastructure, policy is the gate. And in Minnesota, the gates are shifting.

The permitting and policy climate for waste-to-energy systems in the Minneapolis-St. Paul region is no longer defined by technical feasibility or engineering reliability—it is shaped by environmental justice designations, community trust deficits, carbon performance mandates, and institutional risk aversion. While facilities like HERC remain technically compliant, their pathway to expansion or replication is functionally closed.

Current Policy Environment:

- MPCA (Minnesota Pollution Control Agency):

- Mandated to prioritize community impacts, especially in environmental justice (EJ) zones, which include neighborhoods adjacent to HERC and many prospective industrial parcels in the metro.

- New or expanding permits must now address cumulative health impacts, not just facility-specific emissions.

- State Climate Law:

- Minnesota's commitment to 100% clean electricity by 2040 demands low-carbon output at every level—burning waste for electricity no longer enjoys passive regulatory tolerance.

- CHP is only climate-aligned if paired with carbon capture, clean fuels, or dramatic reductions in scope 1 emissions.

- Air Permitting Friction:

- Incinerators and traditional WtE plants require major source permits under Title V of the Clean Air Act.

- Any attempt to site a new mass-burn or RDF facility in the metro would likely trigger multi-year opposition, litigation risk, and political escalation.

- PFAS and Contaminated Waste:

- Growing concern about thermal destruction of PFAS compounds has placed scrutiny on incineration as a disposal method.

- SET's controlled, enclosed gasification process with pre-combustion cleaning offers a more compliant approach to these challenging waste streams.

SET Gasification: A Permitting Advantage

SET's technology offers several unique benefits under this regulatory paradigm:

- Non-incineration classification: Gasification can, in many jurisdictions, qualify as thermal conversion or chemical recycling—avoiding incinerator permit categories altogether.

- Small-site flexibility: Units can be deployed on zoned industrial sites, brownfields, or tribal-managed land with lower permitting thresholds.

- Lower emissions profile: Pre-combustion syngas cleaning and smaller-scale operation reduce pollutant totals below PSD (Prevention of Significant Deterioration) thresholds.

- Policy alignment: Eligible for support through:

- USDA REAP and High Energy Cost Grants,

- DOE Office of Clean Energy Demonstrations,

- EPA Brownfield and Superfund programs,

- Tribal energy self-determination and infrastructure improvement programs.

- Political optics: No stack, no visible flame, no "burn plant." The units look and behave more like shipping containers with diagnostics screens than like legacy waste facilities. This matters—not just for permit approval, but for public trust.

In a state where permitting timelines often determine whether infrastructure is built or shelved, SET's greatest advantage may be that it fits the decade we're in—not the one we left behind.

7. Twin Cities Deployment Opportunities

The strength of a modular system is that it doesn't wait for permission from the center—it goes where the need exists. In the Twin Cities region, that need is no longer concentrated in downtown steam loops or legacy waste corridors. It is spread across industrial edges, transfer stations, contaminated sites, tribal peripheries, and stranded waste streams—places the existing infrastructure was never designed to reach.

Strategic Siting Options for SET in the Twin Cities Region

1. North Minneapolis Industrial Zone

- Context: Adjacent to the existing HERC facility, but increasingly constrained by environmental justice concerns.

- Opportunity: Deploy a SET unit as a parallel system to process feedstocks HERC cannot (PFAS-contaminated waste, agricultural residues).

- Advantage: Lowers HERC load, addresses community opposition by showcasing low-emission alternative, potential local heat reuse for small industry or community centers.

2. Ramsey/Washington County Recycling & Energy Center

- Context: An RDF-based pre-processing facility with growing pressure to move away from out-of-county incineration.

- Opportunity: Co-locate SET units to convert RDF to syngas locally; eliminate transportation emissions; create electricity or fuel on-site.

- Advantage: Aligns with local control initiatives and provides distributed energy capacity without new landfill or incineration.

3. Metropolitan Airports Commission (MAC) Property / MSP Vicinity

- Context: Vast land holdings, rising decarbonization pressures, and regular waste generation from operations.

- Opportunity: Deploy a low-profile SET facility as a sustainability pilot, using syngas to offset auxiliary power or feed into hydrogen blend studies.

- Advantage: Demonstrates aviation-linked carbon innovation in a regulatory-sensitive setting.

4. Suburban Transfer Stations (e.g., Shakopee, Inver Grove Heights)

- Context: Collect and consolidate MSW and bulky waste for out-of-county shipment.

- Opportunity: Add SET units to treat waste locally and generate power or fuels for adjacent municipal use or grid resale.

- Advantage: Reduces waste trucking, builds resilience into suburban infrastructure, and creates local economic nodes.

5. Tribal Land on Urban Periphery (e.g., Shakopee Mdewakanton Sioux Community)

- Context: Sovereign capacity, renewable energy interest, and land control near key logistics corridors.

- Opportunity: Host a demonstration-scale SET system that can serve tribal waste needs while producing hydrogen or methanol for regional sale.

- Advantage: Reinforces energy sovereignty, clean-tech leadership, and distributed circular economy pathways.

6. University of Minnesota / Energy Transition Lab

- Context: Academic leadership in energy, agriculture, and circular economy.

- Opportunity: Use SET platform as a live research and demonstration site for waste-to-hydrogen studies, policy modeling, and lifecycle carbon accounting.

- Advantage: Anchors academic validation, policy outreach, and workforce development.

Deployment Philosophy: Micro-Hubs, Not Mega-Plants

SET units are not replacements for the entire waste management ecosystem. They are surgical deployments—each one solving a specific carbon, waste, or energy problem at the local scale. Deployed at 5–10 tons per day, they support decarbonization without requiring consensus or capital megaprojects. Deployed in clusters, they create redundancy, fuel flexibility, and heat co-location without grid dependency.

In a region increasingly constrained by space, politics, and community fatigue, SET systems offer something increasingly rare in infrastructure: agency.

8. Replication Across Minnesota and the Upper Midwest

The mistake of traditional infrastructure has always been to assume that what works in the center will naturally extend to the edge. The SET Smart Gasification platform inverts that logic: it starts at the edge—and invites the center to catch up.

Minnesota is not merely Minneapolis. It is a fabric of counties, tribal nations, farming corridors, remote communities, and legacy industrial zones—each with its own waste profile, energy need, and infrastructure reality. Centralized incineration cannot serve them. Even district heating can't reach them. But SET can.

Scalable Applications Beyond the Metro

1. Agricultural Processing Corridors (e.g., Crookston, Willmar, Austin)

- Feedstocks: Agricultural waste, animal bedding, spoiled grain, food waste.
- Energy Use: Process heat, grain drying, on-site electricity.
- Opportunity: Deploy 5–15 TPD SET units near processors or co-ops.
- Result: Lowers propane/diesel use, eliminates ag waste transport, decarbonizes rural supply chains.

2. Tribal Nations

- Communities: Red Lake, White Earth, Mille Lacs, and others.

- Sovereign Leverage: Independent permitting, DOE tribal energy funding, waste control.

- Opportunity: Establish SET systems as energy sovereignty platforms—producing hydrogen or methanol for internal use or regional sale.

- Result: Reinforces economic independence, environmental stewardship, and intergovernmental leadership.

3. Rural Counties and Regional Transfer Hubs

- Challenge: Rising tipping fees, longer hauls, and closed landfills.

- Solution: Small SET units co-located with transfer stations or landfills-in-reclamation.

- Revenue: Energy resale, avoided landfill costs, LCFS/45Q participation.

- Impact: Local jobs, decarbonized waste flows, and future-proofed infrastructure.

4. Superfund and Brownfield Redevelopment Sites

- Challenge: Soil contamination, PFAS waste, permitting friction.

- Solution: Deploy SET to treat contaminated feedstocks and co-locate with reclamation efforts.
- Alignment: Strong match with EPA and DOE remediation programs.

5. University Extension & Technical Colleges

- Role: Provide workforce training, policy research, and applied science.
- Sites: UMN Morris, Minnesota State Energy Center of Excellence (Faribault), Itasca Community College.
- Opportunity: Partner as demonstration, validation, and deployment training hubs.

Strategic Implication

SET gasification's strength is not in its size—it's in its reproducibility. Each deployment solves a local problem, but every successful site validates a national network logic: decentralized, carbon-accountable, energy-flexible.

From Grand Forks to Rochester, Duluth to Des Moines, tribal councils to co-op boards—the need is there. What's missing is the model.

This white paper proposes that the Twin Cities metro is not just a deployment site—it's a template. A proving ground for what comes next in waste valorization, energy sovereignty, and infrastructure democratization.

9. Strategic Partners and Stakeholder Alignment

No energy system scales on engineering alone. It scales through alignment—of public policy and private capital, of technical validation and community trust, of land, labor, and law. In the case of the SET Smart Gasification platform, the technology is ready. The question now is: who moves first?

In the Minneapolis-St. Paul region and the greater Midwest, several actors are already positioned—whether they know it or not—to play catalytic roles in deployment, demonstration, and scale.

Local and Regional Government

- Hennepin County / Ramsey County

- Operators of existing WtE and transfer infrastructure.

- Likely partners for co-siting, feedstock supply, or strategic offloading from HERC.

- Early adopters of carbon accounting frameworks that align with SET's lifecycle performance.

- Metropolitan Council

- Regional planning authority for infrastructure, waste, and energy.

- Conduit for inter-jurisdictional deployment planning and grant alignment.

- City of Minneapolis / City of St. Paul

- Climate action plan implementation arms; potential demonstration host cities.

- Political incentive to show alternatives to legacy incineration near EJ-designated neighborhoods.

State and Regulatory Agencies

- Minnesota Pollution Control Agency (MPCA)

- Key permitting authority; SET aligns with state emphasis on modular, lower-emission systems in sensitive areas.

- Potential partner in performance validation or environmental justice engagement.

- Minnesota Department of Commerce – Energy Division

- Supports distributed energy systems and administers clean energy grants.

- May align SET demonstration with broader resilience and decarbonization goals.

Utilities and Energy Stakeholders

- Xcel Energy

- Statewide electric utility with pressure to decarbonize fuel mix.

- Potential off-taker for syngas-derived electricity or hydrogen blending pilot.

- Strong interest in distributed peaker displacement and thermal asset diversification.

- Minnesota Municipal Power Association (MMPA) / Local Electric Co-ops
- Natural fit for behind-the-meter or distributed microgrid demonstration.
- More flexible rate structures and mission-aligned with local energy autonomy.

Academic and Research Institutions

- University of Minnesota – Energy Transition Lab / Humphrey School / West Central Research & Outreach Center
- Provide policy validation, life cycle analysis, and stakeholder modeling.
- Ideal partners for public outreach, workforce development, and policy impact studies.
- Minnesota State Energy Technical Programs (e.g., Itasca, Century, Alexandria Tech)
- Hosts for technician training and industry partnership programs.

Tribal and Sovereign Nations

- Shakopee Mdewakanton Sioux Community (SMSC)
- Resource-rich, sovereign, and energy-forward; potential early partner and demonstration host.
- SET aligns with land-use sovereignty, clean fuels, and sustainable development.

- White Earth, Red Lake, Mille Lacs

- Strong interest in energy independence, waste control, and emissions reduction.

- Can access unique federal funding streams (DOE, BIA, Tribal Energy Loan Guarantee Program).

Private Sector and Financial Partners

- Green infrastructure investors and family offices

- Attracted by distributed energy, carbon-negative profiles, and repeatable deployment model.

- Engineering, Procurement, and Construction (EPC) partners

- Needed for localized deployment, utility integration, and fleet standardization.

- Regional waste haulers and processors

- Feedstock suppliers and potential site hosts for stranded waste conversion.

Strategic Implication

The SET platform does not require a revolution. It requires a quorum—a few visionary partners with aligned incentives to demonstrate that decentralized carbon infrastructure works.

When institutions, agencies, and investors move in concert—even once—they redefine what's realistic. SET's promise is that the infrastructure of the future doesn't need to be scaled up. It needs to be invited in.

10. Conclusion and Strategic Call to Action

The Twin Cities were once a model of energy realism. They understood that waste was not an embarrassment to be hidden but a resource to be used. District energy, CHP, and incineration were not ideological—they were functional. They made the most of what was available and built a reliable, if rigid, system from it.

But function evolves.

What was once waste-to-energy is now carbon arbitrage. What was once heat recovery is now fuel flexibility. What was once centralized and singular must now be distributed, modular, and multi-output—not because the old model failed, but because the world around it changed.

The SET Smart Gasification platform represents more than a new technology. It represents a new grammar of infrastructure: small instead of large, mobile instead of fixed, clean before combustion, and carbon-accountable by design.

We are no longer building systems for cities as they are—we are building for the gaps they've left behind:

- The rural counties exporting their waste and importing their power.
- The tribal nations with the will but not the infrastructure.

- The food processors, brownfield owners, and grid-edge hospitals waiting for something more resilient than rhetoric.

A $10 million demonstration facility—deployed in the Twin Cities metro or just beyond its radius—can prove this model. It can turn stranded waste into hydrogen. It can transform public hesitation into carbon performance. It can anchor the first replicable, policy-aligned, revenue-flexible deployment of SET in the Midwest.

We are not replacing HERC. We are flanking the problem. We are not abandoning district energy. We are completing it. And we are not pitching innovation for its own sake—we are inviting Minnesota to do what it has always done best: show the rest of the country what's possible when pragmatism leads the way.

If you are a policymaker, a planner, an investor, or a community with waste and ambition—this is your infrastructure moment.

Let's begin.

CLOSING PAGE

There comes a point in every document where the words have completed their work, where the architecture stands on its own, and where the reader must step beyond the page and into the implications it quietly arranged. If you have reached this final leaf, then you have moved through atmosphere, intent, interpretation, and evidence in a single, uninterrupted arc. The Prelude opened the corridor. The essay revealed the meaning beneath the chemistry. The white paper confirmed what the narrative had already taught you to expect.

Nothing further needs to be said; everything has been prepared. What remains is the simple acknowledgment that this work was written with purpose, with precision, and with an unwavering respect for the regions and communities at the edge — those who stand ready to shape the future long before the center realizes the century has already arrived.

This page exists as a signature in spirit as well as in ink, the quiet seal of authorship that affirms the integrity of the document and the intention behind it. What comes after this belongs to decision-makers, innovators, sovereign nations, and the cities and counties that have waited long enough for their share of agency.

COLOPHON

This document was conceived, written, and assembled as a unified manuscript blending atmosphere, narrative intelligence, and technical rigor.

The Prelude establishes the interpretive frame.

The cipher essay reveals the meaning behind the transformation.

The white paper provides the structural evidence.

The closing page seals the work as a completed act of authorship.

All writing, arrangement, and strategic framing by:

Joe Cozart, GMJoe™ Consulting.

ABOUT THE AUTHOR

Joe Cozart is the founder of GMJoe™ Consulting, where he brings a uniquely cross-disciplinary intelligence to energy strategy, regional infrastructure, and distributed-systems design. His work spans municipal waste-to-energy, aerospace corridor development, sovereign corridor planning, and the emerging interface between local autonomy and national resilience. Through a combination of technical clarity, narrative precision, and geopolitical awareness, he helps public and private leaders interpret the shifting architecture of energy at the edge—where population, waste streams, regulatory pressure, and stranded capacity converge. His white papers, essays, and strategic briefs are widely regarded for their ability to translate complexity into structural advantage, offering decision-makers a clear path through the noise of evolving policy, PFAS liability, carbon markets, and next-generation thermal technologies. This treatise reflects that same commitment: to elevate the leaders who operate beyond the beltway and to equip them with the clarity required to act with sovereignty.

Made in the USA
Monee, IL
04 November 2026

67387301R10046